TRANSFORMERS CONNECTIONS

Foreword

In the south of Florida and in many other states the distribution of electric energy is performed over overhead lines and it is distributed to the household and small workshops through one-phase transformers. The energy supply may be performed through one-phase or three phase service.

The source is transformer banks of one, two or three one-phase distribution transformer units. This system is very flexible, as it allows starting with one one-phase units and later adding one or two more transformers to supply one-and three phase power at the same time to the customer.

The connection of these transformers cannot be done at random, but polarity of each transformer has to be considered in order to supply a balanced voltage.

In thee one-phase and three-phase energy supply each transformer is not loaded the same, so the load of each transformer has to be taken into account in order to design the transformer bank for the load that it is supposed to carry.

I hope the Amazon format does not disarray and disorder the position of the elements of the formulas and relationships used in the text

Content

1.How a transformer works

1.1 Induced voltage

Suppose we place two magnets in front of one another, one South Pole facing the north pole of the other one. We will have to hold them in place somehow, so they will not be able to attract each other (different poles attract each other, same poles repel each other).

Let's move a piece of wire quickly enough in the gap between the two poles, perpendicular to the magnetic lines that will flow from one pole to the other. If we connect a sensitive enough voltmeter to the ends of the wire, we will notice that every time we move the wire, the pointer of the voltmeter will deflect indicating that there is a difference of potential reading.

The unit of the *difference of potential*, as we electricians know, are usually given in *volt*, and the read intensity will be the *voltage* induced on the wire. The polarity of the voltage reading will be one when the wire is moved downwards and the opposite when the wire is moved upwards, assuming that the magnets are place in a horizontal position. If we move the wire quick enough we will get an *alternating induced voltage.*

If a resistor in connected at the end of the wires, an *electric current* will flow.

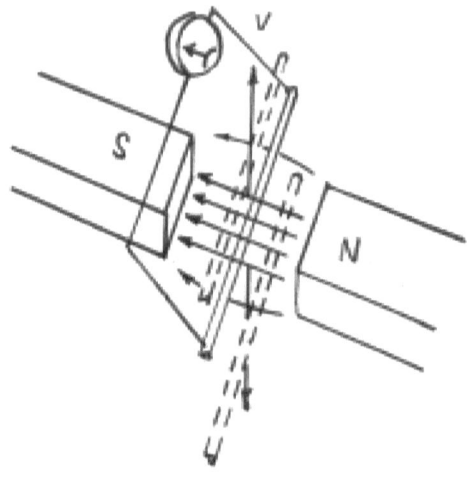

Fig.1.1.1 Perpendicular motion of a wire through magnetic lines

Believe it or not, no scientist can tell why this is so, why voltage is induced every time a wire moves perpendicular to magnetic lines. However, Ohm's law, Maxwell equations, Kirchhoff, The venin and Norton theorems, the whole electricity theory is built on this foundation.

Not only can we can move a piece of wire, we can move a metal bar, a disc, any conductor of any shape and we will obtain the same results. So, we can make this statement general if we say that:

The relative motion of a conductive surface, perpendicular to the magnetic lines, induces a voltage on the conductor.

1.2 Frequency

If we increase the speed of the conductor, we notice that the induced voltage will *increase,* so we can say that the induced voltage is a function *of frequency.* The higher the frequency, the higher the voltage induced on the conductor.

1.3 Magnetic lines intensity

If we use more powerful magnets, the induced voltage will be higher, so we can say that the induced voltage is a function of the *intensity of the magnetic lines.*

1.4 Number of turns

If we add another piece of wire in series with the existing one, we will notice that the induced voltage will be a function of the number of wires or conductors we connect in series. Twice as much voltage is induced on two wires; three times as much voltage is induced on three wires connected in series, etc. So we can state that *the induced voltage is a function of the number of conductors in series, moving perpendicular to the magnetic lines.*

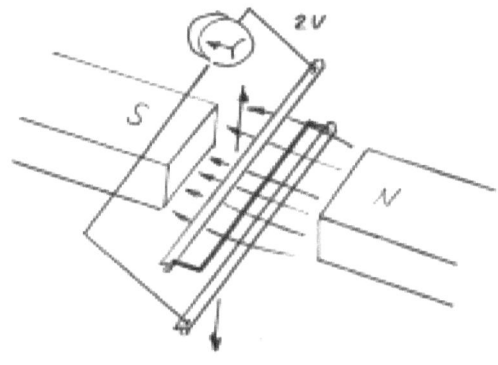

Fig 1.4.1 Double voltage induced adding
a second serial wire

Analyzing the above we can say that the induced voltage is a function of:

1. - frequency

2. - magnetic lines intensity

3. Number of wires connected in series

These are the principles on which all alternating current induction machines of any kind, rotating or steady, work.

All induction machines are designed on the assumption that the alternating current is a *sine wave*. Different wave shapes will give different results, poor performance and overheating. For a more detailed description about how sine wave is generated and represented as a vector see *Reactive Power Management* of same author.

The effect of a non-sine wave on an induction machine is analyzed by means of harmonics. It is the mathematical decomposition of the non-sine wave in different groups of sine-waves of different frequency and amplitude. The effect of each amplitude and frequency is studied separate; a final result is obtained analyzing the different individual waves. The topic of harmonics is beyond the scope of this paper.

In the case of the transformer, the magnetic lines move and the wires are steady. That's why we pointed out that voltage is induced due to the *relative motion* of the wires and the magnetic lines. Either one can move, the other can be motionless.

2. one-phase transformer

The transformer is composed of two coils placed on an iron core. One coil is the one that makes the magnetic lines flow in the core. As the magnetic lines vary according to the alternating current flowing in the winding, the magnetic lines move and reverse polarity systematically along the core. This winding is called the *primary winding*. There is another winding on the core called the *secondary winding* on which the voltage is induced.

The magnetic lines *cut the wires in the coil of the secondary winding* and induce a voltage on this winding proportional with the number of turns.

The number of primary turns divided by the number of secondary turns is called the *transformer ratio*. Primary and secondary voltage will show approximately the same ratio. If the secondary coil has half of the turns of the primary, the secondary voltage will be approximately half of the primary.

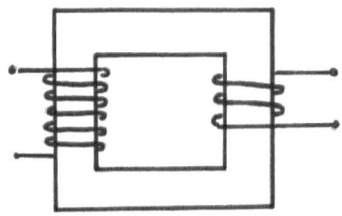

Fig. 2.1 Schematic transformer

We use the term *approximate* because in the real transformer the ratio is slightly different than the turns ratio due to losses, stray magnetic lines etc., but it is a good practical approximation.

Primary and secondary coils are drawn separately in the sketch, but in the real distribution transformer both coils are wound on the same side of the core. The core is assembled making two windows and both primary and secondary coils are wound on the central post to reduce the stray lines that close through the air and reduce the performance of the transformer.

In the practical transformer they may place the coils on top of one another, the lower voltage coil is placed close to the core, the higher voltage coil will be placed on top of the lower voltage coil to keep it as far from grounded core as possible.

The transformer is reversible. We may need either increase, or reduce voltage; any side may be primary or secondary. On the high voltage grid of 13.8 kV and higher, the standard is to take the higher voltage side as primary and the lower voltage side as secondary, but the transformer works either way.

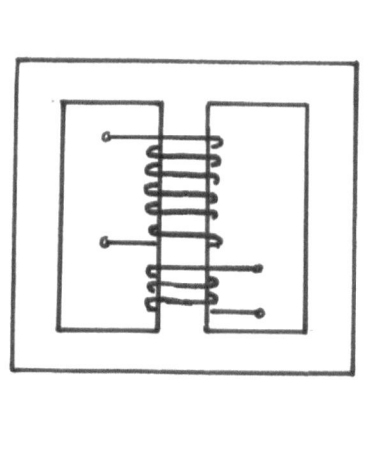

Fig. 2.2 Practical transformer

2.1 Polarity

We are used to the term *polarity* in direct current systems to indicate *positive* and *negative* in the electric system. The one-phase distribution transformer also has polarity, in this case it does not indicate positive or negative, because the alternating current reverses polarity 60 times per second in the US (50 times in Europe and some South American countries).

Polarity in the case of the one-phase transformer relates to whether the secondary voltage *adds to*, or *subtracts from* the primary voltage. Following manufacturing convenience and standards, the primary and secondary turns may be wound in the same direction or in opposite direction. Using an arrow to represent polarity we can determine whether the secondary will add or subtract from the primary.

Polarity is usually found on the nameplate as part of the technical information of the transformer. If this information is missing or in doubt, we can determine the polarity of the transformer as follows.

Suppose a one-phase transformer with a transformation ratio of 10. If we impose 100 volt on the primary, we should obtain 10 volt on the secondary coil. If the polarity is *subtractive*, a voltmeter connected as shown in fig. 2.1.1 will read $100 - 10 = 90$ V.

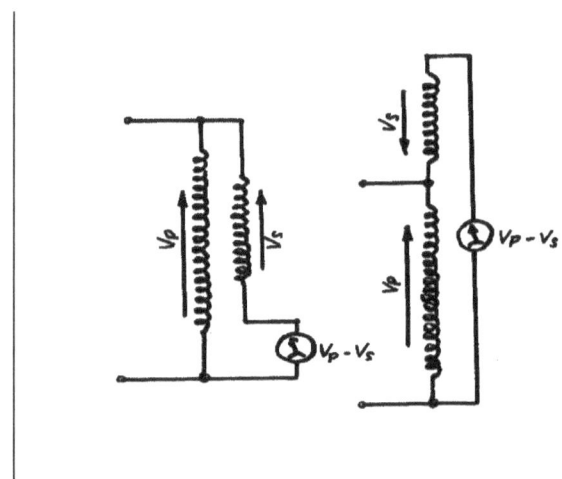

Fig. 2.1.1   Subtractive polarity

If the coils are wound in opposite direction, as shown in fig. 2.1.2, the polarity will be *additive* and the voltmeter should read 100 + 10 = 110 V.

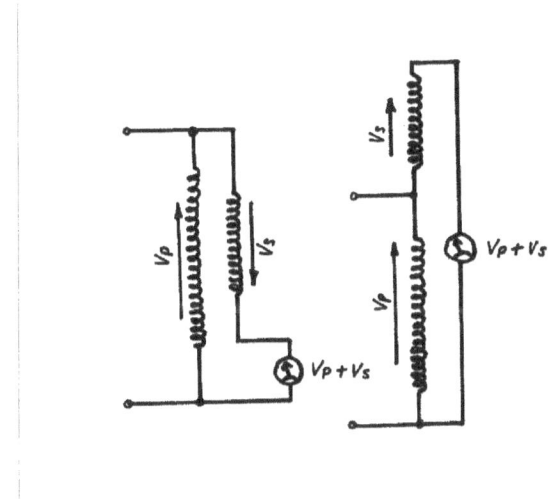

Fig 2.1.2   Additive polarity

The arrows on both additive and subtractive can be considered can be considered as vectors showing the relative position of primary and secondary voltage. A schematic diagram showing the coils placed in straight line helps to understand how the voltage sum or difference is obtained. For vector representation of a sine wave see *Reactive Power Management* of same author.

2.2 Transformer taps.

Taps are additional, extra connections provided by the manufacturer that allow different ratios by changing the number of turns, this way the manufacturer provides the possibility of voltage regulation in the distribution transformer and some industrial transformers of smaller size and capacity.

Let's take the transformer of ratio 100/10 used as an example. Suppose the primary coil has 100 turns, the secondary coil has 10 turns. Suppose this symbolic transformer is supplying a power of 100 VA. The current flowing in the primary will be: 100 VA/100 V = 1 Ampere.

The current flowing in the secondary will be: 100 VA/10 V = 10 A. Taps can be taken on either side of the transformer, however, from the manufacturer point of view it is more convenient to take the taps where the cross section of the current is smaller and the wire is smaller too, this makes the wires easier to handle to conform the taps and get less heat dissipation on the contact resistance of the tap changer.

Suppose we have taps at 120%, 110% 100%, 90%, and 80% of the primary coil. 120% means in this case 20 turns, 110% means 10 turns above the nominal number of turns. On the other hand, 90% means 10 turns less, 80% means 20 turns less bellow the nominal number of turns. Fig 2.2.1 shows a schematic tap-changer on the primary side.

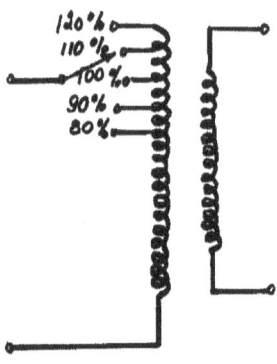

Fig 2.2.1 Schematic tap-changer on primary side

If the primary voltage rises to 120 V we need to use the 120% tap to keep the secondary voltage at the same level. The new ratio will be 120 turns/10 turns = 12. The secondary voltage will be 120V/12 = 10 V. It means in case of voltage higher that nominal we have to add turns to reduce the ratio. On the contrary, if the primary voltage drops to 90 V and we use 90% tap, the new ratio will be 90 turns/10 turns = 9, the secondary voltage will be 90 V/9 = 10 V. If the primary voltage drops, we have to reduce turns to reduce the ratio if we want to keep the secondary voltage at the same level.

In the distribution transformer used by the companies of electric supply in the US the tap change must be performed with the transformer de-energized. There is a lid on top of the transformer tank that gives access to the tap changer. Some transformers of smaller size and capacity also have taps to adjust the voltage output on the secondary side.

2.3 Transformer losses.

There are losses in the core and losses in the wire. The alternating variation of the magnetic lines in the core induces voltage not only in the wires of the secondary coil, but in the core itself. As the core is metal, it does like it is in short circuit and this induced voltage makes current flow in the core. These currents are called *eddy currents.* If the core was a block of metal, the circulating eddy currents would produce such a heat such, that the core would melt. This the principle on which the induction furnace works. The metal is melted by the heat raised in the core by the eddy currents circulating in the block of metal.   To avoid the intensive heat, the core of the transformer is made of thin sheets to split the eddy currents. These sheets are isolated from one another to limit the eddy current of each sheet, making it flow only in that sheet. Same process occurs in induction motors and every induction device, therefore, the core of all induction machines is laminated, composed of many thin sheets isolated from each other. The metal used in the sheets is a special alloy that offers little resistance to the magnetic lines.

The losses in the core are called *core losses* or *iron losses.* These losses are always present during the time the transformer is in service, that's why the losses of transformers that are going to be permanently connected to the electric network are kept as low as possible.

The flow of the electric current in the coils produces so called *copper losses.* This term might be misused, as some manufacturers use aluminum wire to reduce manufacturing costs. Some manufacturers use sheets of copper or aluminum instead of wire to reduce the losses in the winding. The losses in the primary and secondary coils are proportional to the square of the current multiplied by the resistance of the coil, so these losses depend on the load in the transformer.

Losses might be an economic factor to be considered. A more expensive transformer with low losses might result more economic on the long run, depending on how many hours the transformer remain connected to the grid and how it will be loaded.

2.3 Transformer impedance

Transformer impedance is important because it determines voltage drop in the transformer and how the transformer opposes to short circuit.

Following ohms law se can state that the impedance of the transformer will be

Z_T (ohm) = volt/ampere
2.3.2

The distribution transformer primary voltage is given in phase-to-phase kilovolt and the capacity is given in kilovolt-ampere (kVA). The distribution transformer is usually connected between line and ground. It is well known fact that the line-to-phase voltage will be line-to-line voltage/1.73. It is out of the scope of this book to prove it, so we will just accept it.

The transformer is composed of two coils composed of many turns, so the reactance of the transformer is usually higher than the resistance, that is why in many textbooks they talk about the *reactance* of the transformer.

The impedance of the transformer, usually one-phase and three-phase transformer is given in per cent of the nominal or nameplate impedance, so:

$Z_\% = (Z_{actual}/Z_{nameplate}) \times 100$
2.3.3

$Z_{nameplate} = V_{nameplate}/I_{nameplate}$
2.3.4

$$Power_{nameplate} = V_{nameplate} \times I_{nameplate}$$
2.3.5

Thererefore

$$I_{nameplate} = Power_{nameplate} / V_{nameplate}$$
2.3.6

Modifying 2.3.3 by multiplying by Z$_{nameplate}$ in both sides we have

$$Z_{actual} = 100.Z\% \times Z_{nameplate}$$
2.3.8

Substituting 2.3.6 in 2.3.4

$$Z_{nameplate} = V^2_{nameplate} / Power_{nameplate}$$
2.3.9

The per cent impedance, the voltage and the power or capacity of the transformer are nameplate values.

The nameplate voltage is usually given in kilovolt (kV) and the power or capacity of the transformer is usually given in kilovolt-ampere (kVA).

Let's determine the actual ohm impedance of the 10 kVA transformer that is going to be installed on a 13.8 kV feeder. The one-phase transformer will be connected phase-to-ground, therefore we have to divide 13.8 kV by 1.73. We have to convert all units to V and VA first. Using 2.3.9,

$Z_{nameplate}$ = $[(13.8/1.73)\times1000]^2$ / $10\times1000 = 7,977^2/10,000 = 6,362$ ohm

The percent impedance on the nameplate is 2.5%, therefore, the actual ohm value of the impedance of this transformer will be:

$(2.5/100)\times 6362 = 159$ ohm

How can we determine the percent impedance of a transformer?

Short-circuit the secondary winding, connect the primary high-voltage side to a variable, low voltage source. Increase the voltage until nameplate current is reached, the $Z\% = (V_{actual} / V_{nameplate})$x100 because

$Z_{ohm} = V_{read} / I_{nameplate}$ but applying 2.3.3

$Z\% = [(V_{read} / I_{nameplate}) / (V_{nameplate} / I_{nameplate})]$x100

To divide by a fraction we have to multiply by the inverse of the fraction in the denominator, therefore,

$$Z\% = (V_{read} / V_{nameplate})\text{x}100$$

In order to avoid voltage fluctuations due to customer load change, the distribution transformers are designed to have as low percent impedance as possible. Together with low no-load losses, low impedance is an important requirement for the distribution transformer.

3. Load supply using one-phase transformer banks

In the South of Florida the load distribution is very flexible. Distribution of electric energy is supplied mainly through overhead lines at 13.8 kV. The overhead distribution is performed running the phase wires on cross-arms placed on top of concrete or wooden poles treated with a substance that delays the wood decay. Wood is cheaper and easier to handle, concrete is more durable. Sometimes the cross-arm arrangement is not used and the phase lines are sitting directly on isolators placed on the poles, keeping the standard phase distance for the voltage level.

There is an underground distribution too, it is found in big cities where space and reliability are priority. In this case the distribution is performed through high tension underground cables and transformers are place in an underground vault.

Although all the three phases are usually run along the feeder, the distribution of electric energy is performed using one-phase transformers for household load. This distribution arrangement gives flexibility to the distribution.

In areas with light one-phase load, only one phase can be run to supply the area. In the case light three-phase load is demanded, a second phase is run to supply one- and three-phase load adding a second transformer. Three-phase load supply with two one-phase transformer will be discussed in more detail later.

The three-phase transformer's secondary at the substation is usually a grounded wye connection. The ground wire is run all along the distribution feeder, together with the phase wires, and is grounded at every pole or every two or three poles along the feeder. The grounding of the wye at the substation makes every phase independent, and the one-phase distribution transformers are placed along the feeder, as close to the load as possible.

The distribution phase-to-phase voltage in the

South of Florida is 13,800 V between phases and 13,800/1.73 = 7,977 V between each phase and ground. As each one-phase transformer is connected between phase and ground, care must be taken to distribute the one-phase load between the three phases in order to keep the three phases as evenly loaded as possible.

In order to balance the voltages at the substation as much as possible, the primary of the transformer is usually connected in delta or triangle, this way the current flow inside the triangle will balance somewhat the secondary voltage.

As the phases are not always evenly loaded, and it varies with the time of the day, individual voltage regulators are used per phase at the substation to keep an even voltage level on the feeder.

Fig 3.1 shows a scheme of the distribution system. The distribution transformers are represented with a single coil to make the scheme simpler. Triangle/wye system on the left represents the three-phase transformer at the substation.

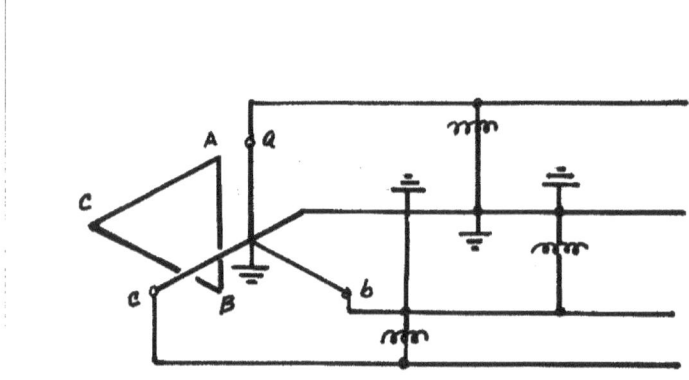

Fig. 3.1 Schematic one phase
distribution

The one-phase distribution secondary
has a middle tap that divides the coil in
two halves. The voltage between each
end and the middle tap is half of the
voltage between the ends of the coil.
Voltage between ends will be 240 V and
120 V between each end and the middle
tap. Fig. 3.2a shows a schematic one-
phase distribution transformer.

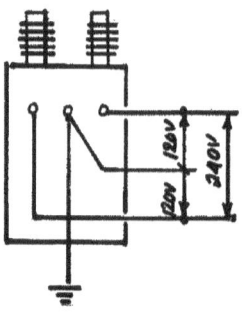

Fig. 3.2a Schematic one-phase
distribution transformer

If the household requires only 120 V
supply, one end and the middle tap is run
into the household, the two ends are
usually marked as phase A and Phase B.
To prevent voltage shift the middle tap is
grounded. As V_{AN} + V_{BN} = V_{AB}, if load on
one half is heavier than at the other half,
the voltage drop in the heavily loaded

half will be greater making the voltage on the other half rise.

Grounding the center tap makes both halves more independent and the customer receives a better voltage quality. Fig. 3.2 b shows a real picture of a distribution transformer. The low voltage distribution of the load is accomplish through horizontal wires to which the secondary bushings of the transformer are connected. The top wire is the ground wire, second wire is usually a phase and third wire is B phase, in that order. If there is a third phase C for three-phase load, it is placed at the bottom. Note that the center tap is connected to the top wire that should be grounded at the pole.

Fig. 3.2b Distribution of secondary load
using horizontal wires.

Sometimes the distribution of the load is
accomplish through cables, and not
through horizontal wires, as shown in fig
3.2c

Fig. 3.2c Secondary load distribution using cable.

3.1 Banks composed of three one-phase transformers

Transformer banks can be composed of one, two or three units, supplying one-phase, three-phase load or both. One transformer unit is used to supply one phase load to household. A combination of one-and three-phase load can be supplied with banks composed of two or three units.

The three units can be connected in wye on the primary and wye on the secondary, we will call this connection *wye/wye*. Primary may be connected in wye and secondary in triangle or delta. We call this *wye/delta* connection, or they can be connected triangle on the primary and triangle on the secondary and we call this *delta/delta* connection. The winding of the transformers will not be sketched, only the position of the voltage vectors. Primary voltage is indicated as A, B and C with capital letters, while secondary voltage is indicated as a, b and c in low case. Primary and secondary voltage will point in the same direction if polarity of transformers is substractive in both wye and triangle connection.

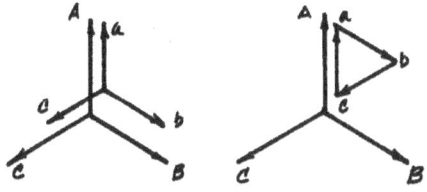

Fig 3.1.1 Primary and secondary voltage position for three substractive polarity units

If we use additive transformers the vectors will point in opposite direction, it makes no difference in case of load supply as long as the e clockwise sequence of the voltage is the same.

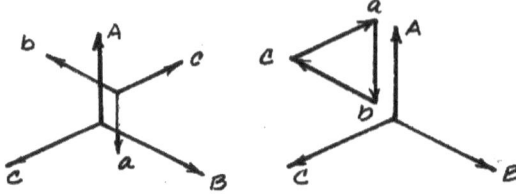

Fig 3.1.2 Primary and secondary voltage position for three additive polarity units.

Same case in the triangle/triangle connection. Subtractive polarity will show primary and secondary voltage in the same direction, while additive polarity will show primary and secondary voltage in opposite direction.

The first thing to consider when installing a bank composed of three units is polarity. We saw in 2.1 that the polarity of the transformer may be *additive* or *subtractive.* If polarity of all three units is the same, there is little concern about how to connect the three transformers. If one of the three units has a different polarity then we have to stop and consider how to connect the three units. If the units are not connected right regarding polarity, one of the voltages will be reversed and the fuses will blow making impossible the put in commission of the bank or some damage may be caused to the devices connected to the bank.

It is very easy to determine how the transformers have to be connected if we make a sketch of the three units indicating polarity drawing arrows alongside the winding. Figure 3.1.3 shows a wye/wye connection.

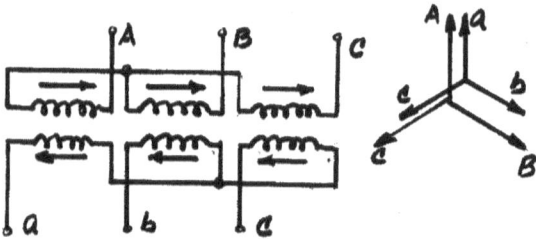

Fig. 3.1.3 Wye/wye connection with additive units

Note that we only have to connect all the tips and all the bottom of the arrows together. If one of the transformers burn and has to be replaced, keeping the same lead arrangement will cause the voltage distortion showed in fig. 3.1.4 for the wye/wye connection.

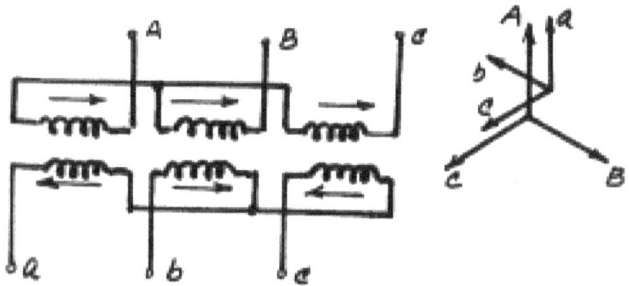

Fig 3.1.4 Voltage distortion caused by reverse polarity on transformer bc

In order to correct the distortion and make the voltage system symmetrical, we only have to reverse the connecting leads following the arrows, as shown in fig 3.1.5. In case of different polarity either lead, primary or secondary may be reversed as convenient.

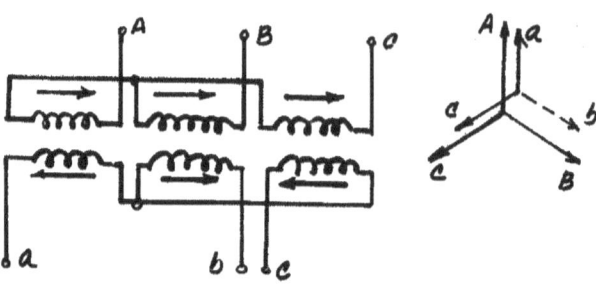

Fig 3.1.5 Voltage correction by reversing connection on transformer b

If we draw all the arrows in the same direction on the secondary side and reverse the arrow on the primary side according to substractive polarity, we will note that either the primary or the secondary connection can be reverse to make up for polarity difference. The important fact is the *relative* position of the arrows on one transformer.

In the connection wye/triangle the voltage distortion due to reverse polarity shown in fig 3.1.6 may damage one or more transformers due to a large current flowing inside the closed triangle. Same thing happens in the triangle/triangle connection. Note in Fig. 3.1.6 that voltage V_{bc} on the secondary has grown to roughly twice the normal line voltage, this will cause a high current flow inside the triangle that may damage one or more transformers.

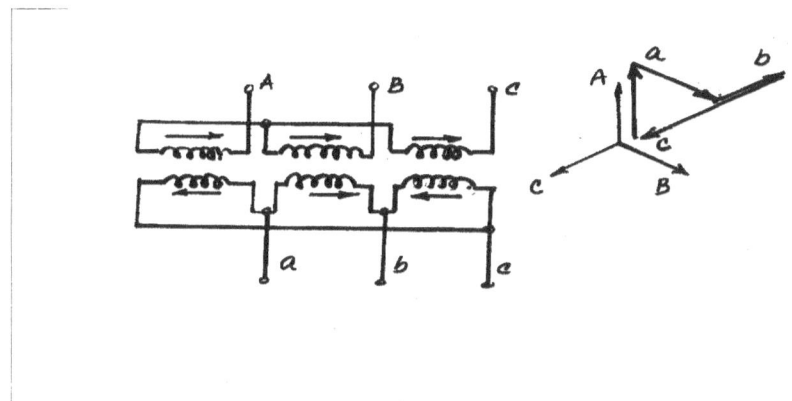

Fig 3.1.6 Voltage distortion caused by reverse polarity on transformer ab

For this connection too it is enough to reverse the connecting leads, either at the side of the load or at the side of the source, to correct the distortion and make the voltage system symmetrical, as shown in Fig. 3.1.7.

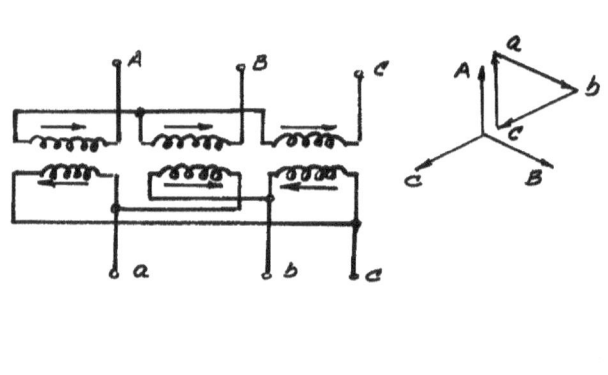

Fig. 3.1.7 Voltage correction by reversing connection of transformer ab

In some cases it is more convenient to leave the secondary lead arrangement and reverse the primary connection leads. Note that the position of the arrows shows the *relative* position of primary and secondary voltage. Any arrow may be drawn in any direction as long as we draw them in the right relative position: same direction for substractive, opposite direction for additive polarity.

In either case the reasoning is very simple: *Tip of every arrow connects to bottom on the next one.*

3.1.1 Supply of one-phase and three phase load

One-phase load may be supplied using one single distribution transformer. What if there is a demand for three-phase load too? Most of smaller customers have one-phase and three-phase devices that might work at the same time. The man working at the machine moved by a three-phase induction motor may be listening to the radio and running a one-phase fan to cool itself. This applies to small workshops and factories. A large mansion may need three-phase and one-phase supply at the same time.

Anyway, if the three-phase load is considerable, a bank of three one-phase transformer units is installed to supply both one-phase and three-phase load.

The most popular transformer bank is composed of three one-phase transformers connected in wye/triangle. One of the transformers is taken to carry the one-phase load; however, as it forms part of the triangle, it must take part in the three-phase load as well.

Usually, the power factor of the three-phase load is different than that of the one-phase load, so the currents add up as vectors in all the transformers. As load may vary during the day due to devices in-and out of service, an approximation is used to determine how the one-and three-phase load combine in the transformers.

Most of the time we do not know exactly what demand the bank will have, the demand is estimated on the basis of the nominal power, characteristic and usage of the load that will require electric energy.

The probable load will be assumed based on a basis of the inventory of devices, corrections factors and basically, common sense. This will determine the capacity and connection of the transformer to be installed. Some electric supply companies have a standard about what type of bank and what capacity according to the load inventory.

We saw earlier that the one phase transformer for household load is a transformer with the 240 V secondary, split in two halves of 120 V by a central bushing. The central bushing is connected to ground to avoid voltage shifting and the usual customer received one phase at one end of the transformer tank and the middle tap.

If the customer requires 240V, he receives phases A and B supplying 240 V service plus the ground wire to supply 120 V load. In that case the whole household load should be distributed between each the phase and the ground wire.

Fig. 3.1.8 shows the relationship between the currents inside and outside the triangle. Using Kirchhoff's theorem we can say the all currents coming into, and going out a nod is zero. Suppose that the current that arrives is positive and the one that is out is negative. In this case we can write:

$$+ I_{ab} - I_A - I_{ca} = 0$$

Arranging the relationship above we get:

$$I_A = I_{ab} - I_{ca}$$

So we have to *add* the reversed vector I_{ca} to I_{ab} . To subtract a vector from other vector *add* the vector to be subtracted in reverse position. It applies to any vector in any field.

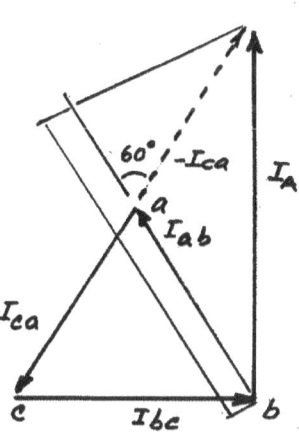

Fig. 3.1.8 Phase-line current relationship
in triangle connection

As a homework, using the vector arrangement shown in figure 3.1.8 and assuming that load is balanced and $I_{ab} = I_{bc} = I_{ca}$, prove that

$$I_A = \sqrt{3}\, I_{ac}$$

It means that the current flowing inside the triangle is 1.73 times smaller than the current I_A flowing in the line. One more interesting feature is that the position of the line current and the current flowing inside the triangle is the same for a given angle (power factor).

3.1.1 Supply of one and three-phase load combined.

Now let's take the secondary delta or triangle connection to supply one-and three phase load.

The power factor of the one phase load may not the same as that of the three phase load, that is, the phase angle is not the same for both loads in this case both currents combine as vectors inside the transformers.

There are so many uncertain factors as to what the load will be, what composition, what percent three-phase, what percent one-phase, how it is going to be used, how frequently and so many uncertainties, that it is better to forget about vectors. Load determined using other assumptions will give us slightly higher results, what is good because it puts us on the safe side.

Let's take the one-phase load first. Fig. 3.1.9 shows how the one-phase load is divided between the three transformers. For the one phase load, transformers AC and BC appear like they are connected in series and at the same time connected in parallel with the transformer to which the one phase load is connected.

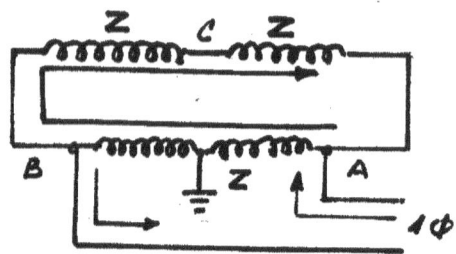

Fig. 3.1.9 One phase load distribution between the three transformers in the triangle.

Suppose the three transformers have the same or similar impedance. The current flowing in the transformer AB with the grounded tap will be:

$$I_{AB} = 2Z/3Z \,.\, I_A = 2/3 \,.\, I_A$$

The one-phase load share taken by the other two transformers (AC, BC) will be:

$$I_B = I_C = Z/3Z \cdot I_A = 1/3\, I_A$$

The transformer AB will carry 2/3 of the one phase load, the other two transformers, BC-AC, will take one third of the one- phase load connected to the bank. Note that the two additional transformers carry *the same third of the one-phase load both of them*, so there is no contradiction.

The three-phase load is evenly distributed between the three transformers, so each of them takes an equal share of the three-phase load.

The total load carried by the transformer with the grounded tap will be: 2/3 of one - phase load + 1/3 of three-phase load.

The total load carried by the other two transformers will be:

1/3 one phase load + 1/3 three phase load

Usually the one phase load is greater than the three phase load. The capacity of the three transformers must not be the same. If the one-phase load is substantially greater than the three phase load, the transformer with the grounded tap could have a higher capacity then the other two.

In this type of transformer bank the center of the wye CANNOT be grounded, it will be explained later why, when we analyze the open delta or open triangle connection.

Fig 3.1.10 shows a three phase transformer bank for combined load. Note that the size of the transformer with the grounded center tap is larger than that of the other two.

Fig. 3.1.10 Three-phase transformer bank for one-and three phase load combined.

3.2 Banks composed of two one phase units for one-and three phase load combined.

The banks composed of two one-phase units are called *open delta* or *open triangle*.

In this case also one transformer with center grounded tap is used and a second, usually smaller transformer, helps in supplying the three-phase load. How can we supply three-phase load with only two transformers?

The answer is in Fig 3.2.1

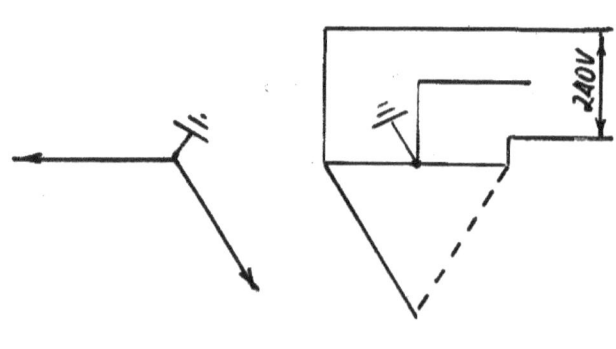

Fig. 3.2.1 Connection in open triangle

The third voltage is a *virtual* voltage between the two open sides of the triangle. In this case, on the contrary to wye/triangle connection, the grounding of the center of the open wye is necessary to obtain the 60 degrees angle to create the virtual voltage for the three phase service. If the wye/triangle service is grounded and one of the three transformers is missing for some reason, for example, a fuse is burnt, the remaining two transformers keep carrying full load and may overheat.

3.2 1 Supply of one and three phase load combined

On the contrary to the three transformer wye/triangle connection, in the open delta every transformer carries the full current. The transformer with the grounded tap will carry full one-phase load and 58% of the three phase load. Each transformer can carry its capacity to avoid overloading.

The available three phase capacity for this type of bank without overloading is the capacity of the smallest one, times 1.73. The relation of each transformer to the full three phase capacity will be: kVA/(1.73. KVA) = 0.58. The one phase load does not flow in the additional transformer, only in the one with the grounded middle tap, however, this transformer must contribute 58% of his capacity to the three-phase load. The capacity of the transformer with the grounded middle tap will be then:

  kVA = Full one phase kVA + 58% of three phase kVA

The amount of three phase load that this connection type can take is limited. Excessive three-and one phase load in the open triangle can lead to voltage distortion due to voltage drop in each transformer. On the contrary of the wye/triangle, vector addition of one phase and three phase may cause a distortion in the resultant vector voltage system. Every electric supply company determines the permissible three-and one phase load that can be safely carried in this type of connection to avoid inconvenient voltage distortion that might affect the performance of induction motors.

Fig. 3.2 shows an open triangle bank in real life.

Fig. 3.2 Open triangle transformer bank

4. Distribution transformer overload

Most of the household load is concentrated in the so call peak time or peak load. Peak load usually starts around 7.00 PM. and ends around 11.00 PM. It varies depending on the area and the time of the year. In winter it starts earlier, in summer it starts later. The hour shift in summer is primarily intended to move the peak time and make the urban customers to turn lights, using cooking- and household devices earlier, contributing to the flattening of the peak load.

The usual small household load lasts between 4 and 5 hours, the rest of the day the distribution transformer will carry very small load. In order to make the most of the transformer, they are designed with overloading capability in mind.

We have seen that the distribution transformer is encased in a tank, usually filled with special transformer oil. The purpose of this oil is not only isolating the high voltage coil, but to cool the transformer at the same time. On account of the heat, the hotter transformer oil will flow upwards and will come down in a slow stream as it cools.

The hotter the transformer, the faster will be the oil stream in the tank. Night time is favorable for cooling, as the heat of the sun is gone and only the heat created by the load stays.

Favoring a better use of the distribution transformer, the manufacturer makes cooling charts. These cooling charts show the permissible overload, depending on the outside temperature and how long the overload will be sustained. So, the transformer can be intelligently overloaded without damaging any of its components because it is considered in its design, making possible the installation of smaller, less costly transformer for loads of smaller duration.

Before intentionally overloading, a basic study must be previously done to get the most of the overloading capability of the distribution transformer without causing damage.5. Distribution transformer per cent impedance

The coils of the transformer have impedance, this internal impedance raises a voltage drop inside the transformer. As the impedance is fix, the voltage drops varies with the load. There are two requirements for the distribution transformer: lower losses and low impedance.

Any transformer has two impedance values, the nominal or nameplate one and the real one. The nominal impedance is computed using nameplate voltage and current. If any of these vary, the impedance of the transformer varies.

5. Transformer impedance

So, how do we determine the actual impedance? The impedance is given in per cent and is usually on the nameplate of the transformer. How can we

determine the real ohm impedance?
Take the illustration in Fig 5.1.

——

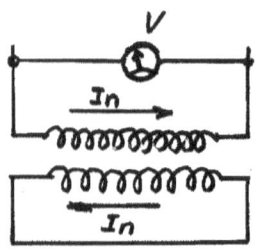

Fig. 5.1 Arrangement to determine actual
impedance,

Let's short-circuit the secondary winding.
Let's connect a variable voltage source
of the right capacity and let's rise the
voltage until nameplate current flows
both in primary and secondary winding.
The real ohm impedance will be *the
present voltage on the meter divided by*

the nameplate voltage of the transformer. We will call this voltage the meter voltage (Vmeter).

Why is this so? The percentage impedance is given taking the nameplate impedance as a basis:

$$Z_n = V_n / I_n \qquad 4.1$$

where Z_n , I_n , V_n are the nameplate impedance, current and voltage respectively. The real ohm impedance will be

$$Z_{ohm} \qquad = \qquad V_{meter}/I_n$$
4.2

Percent impedance will be (Z_{ohm} /Z_n).100

Dividing 4.2/4.2, I_n is the same and will cancel, so we get

$$Z\% =(V_{meter}/V_n).100 \qquad 4.3$$

Distribution transformers usually have low per cent impedance, around 2.5%, big three phase substation transformers have a higher per cent impedance, usually around 5%

6. Distribution transformer protections

Distribution transformers are usually protected by fuses. The fuse used in the distribution transformer is locked in a capsule attached to t long wire. This fuse is placed on the primary side, inside a special pipe designed to extinguish the arc inside when the fuse blows. The long wire is fixed outside on a nut through a spring.

When the fuse blows out, the spring tension is released and the pipe drops isolating the top of the device that is connected to the high tension line. This device is called *drop-out* device and shows that the fuse is blown. The lineman removes the drop out mechanism with an isolated rod, gives tension to the spring when installing the new fuse in place, lifts the drop-out device with the new fuse into position and closes the drop-out device with the rod.

There are different kind of fuses that have a time delay and allow co-ordination with internal protections at the customer, like fuses and magnetic breakers with overcurrent protection.

Fig. 61. Drop-out device on primary side

The distribution transformer fuse can be coordinated with the internal protection so that the internal protection disconnect first for the same overcurrent. In rural areas where fuse protection is widely used sometimes two fuses have to be coordinated so the subordinated fuse may blow first. The time versus current is plotted and a set of curves given by the fuse manufacturer for the different time-delay fuses.

The second protective device used at distribution transformers is *lightning arrester* The lightning arrester is placed as close to the transformer as possible. The task of the lightning arrester is to chop the overvoltage wave produced by lightning strike.

Lightning looks for a place where he can discharge to ground. If lighting discharge current is multiplied by ground resistance value, quite high overvoltage and arise for a very short time, for microseconds or miliseconds. Such a short interval is not enough to generate heat, but it may puncture the isolation of the transformer.

A higher discharge may occur through the small puncture later and make the transformer useless. The lightning current may be so intense, that may make the lightning arrester blow.

Fig. 6.2 Lightning arrester attached to a distribution transformer

A low ground resistance helps not only to avoid voltage shifting on the secondary, low voltage network, but to facilitate the path for the lighting current and reduce the possibility of overvoltage in the distribution system. It may be harmful, not only to the distribution transformer, but to the customers as well. Figure 6.2 shows a lightning arrester next to a distribution transformer.

You can leave your opinion and/or suggestion in a review, at http://reactivepower.blogspot.com, or write to rafbarr45@gmail.com.

If you considered what you read was useful, you can recommend it to others interested in the same topic.